10	11 IB	12 IIB	13 IIIA	14 IVA	15 VA	16 VIA	17 VIIA	18 VIIIA
								4.003 — 2.37 **He** — 2 Helium
			10.81 3 0.80 **B** 2.0 5 Boron	12.011 4 1.09 **C** 2.5 6 Carbon	14.01 3,5 1.40 **N** 3.1 7 Nitrogen	16.00 2 1.31 **O** 3.5 8 Oxygen	19.00 1 1.68 **F** 4.1 9 Fluorine	20.18 — 2.08 **Ne** — 10 Neon
			26.98 3 0.58 **Al** 1.5 13 Aluminum	28.09 4 0.79 **Si** 1.8 14 Silicon	30.97 3,5 1.01 **P** 2.1 15 Phosphorus	32.06 2,4,6 1.00 **S** 2.4 16 Sulphur	35.45 1 1.25 **Cl** 2.9 17 Chlorine	39.95 — 1.52 **Ar** — 18 Argon
58.71 2,3 0.74 **Ni** 1.8 28 Nickel	63.55 1,2 0.74 **Cu** 1.8 29 Copper	65.38 2 0.91 **Zn** 1.7 30 Zinc	69.72 3 0.58 **Ga** 1.8 31 Gallium	72.59 4 0.76 **Ge** 2.0 32 Germanium	74.92 3,5 0.94 **As** 2.2 33 Arsenic	78.96 2,4,6 0.94 **Se** 2.5 34 Selenium	79.91 1 1.14 **Br** 2.8 35 Bromine	83.80 — 1.35 **Kr** — 36 Krypton
106.4 2,4 0.81 **Pd** 1.4 46 Palladium	107.9 1 0.73 **Ag** 1.4 47 Silver	112.4 2 0.87 **Cd** 1.5 48 Cadmium	114.8 3 0.56 **In** 1.5 49 Indium	118.7 2,4 0.71 **Sn** 1.7 50 Tin	121.8 3,5 0.83 **Sb** 1.8 51 Antimony	127.6 2,4,6 0.87 **Te** 2.0 52 Tellurium	126.9 1 1.01 **I** 2.2 53 Iodine	131.3 — 1.11 **Xe** — 54 Xenon
195.1 2,4 0.87 **Pt** 1.5 78 Platinum	197.0 1,3 0.89 **Au** 1.4 79 Gold	200.6 1,2 1.01 **Hg** 1.5 80 Mercury	204.4 1,3 0.59 **Tl** 1.5 81 Thallium	207.2 2,4 0.72 **Pb** 1.6 82 Lead	209.0 3,5 0.70 **Bi** 1.7 83 Bismuth	(209) 2,4 0.81 **Po** 1.8 84 Polonium	(210) 1 — **At** 2.0 85 Astatine	(222) — 1.04 **Rn** — 86 Radon
							Halogens	Noble Gases

157.3 3 0.59 **Gd** 1.1 64 Gadolinium	158.9 3,4 0.56 **Tb** 1.2 65 Terbium	162.5 3 0.57 **Dy** — 66 Dysprosium	164.9 3 0.58 **Ho** 1.2 67 Holmium	167.3 3 0.59 **Er** 1.2 68 Erbium	168.9 2,3 0.60 **Tm** 1.2 69 Thulium	173.0 2,3 0.60 **Yb** 1.1 70 Ytterbium
(247) 3 0.58 **Cm** — 96 Curium	(247) 3,4 0.60 **Bk** — 97 Berkelium	(251) — 0.61 **Cf** — 98 Californium	(252) — 0.62 **Es** — 99 Einsteinium	(257) — 0.63 **Fm** — 100 [illegible]ium	(258) — 0.64 **Md** — 101 Mendelevium	(259) — 0.64 **No** — 102 Nobelium

- metals
- nonmetals
- noble g[illegible]